JUN - - 2000	DATE DUE	

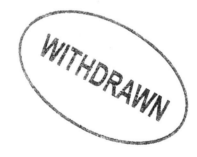

Oceans

Heather C. Hudak

WEIGL PUBLISHERS INC.

Published by Weigl Publishers Inc.
350 5th Avenue, Suite 3304, PMB 6G
New York, NY 10118-0069
USA

Web site: www.weigl.com

Library of Congress Cataloging-in-Publication Data

Hudak, Heather C., 1975–
    Oceans / Heather C. Hudak.
    p. cm. — (Biomes)
    Includes index.
    ISBN 1-59036-348-5
(hard cover : alk. paper) — ISBN 1-59036-354-X (soft cover : alk. paper)
1. Marine ecology—Juvenile literature. I. Title. II. Biomes (Weigl Publishers)
QH541.5.S3H83 2006   577.7—dc22   2005004393

Printed in the United States of America
1 2 3 4 5 6 7 8 9 0   09 08 07 06 05

c.2

Project Coordinators Heather
C. Hudak, Heather Kissock

Substantive Editor Tina
Schwartzenberger

Copy Editor Heather Kissock

Designers Warren Clark,
Janine Vangool

Photo Researchers Heather
C. Hudak, Kim Winiski

Photograph Credits
Every reasonable effort has been made to trace ownership and to obtain permission to reprint
copyright material. The publishers would be pleased to have any errors or omissions brought
to their attention so that they may be corrected in subsequent printings.

Cover: Getty Images/Chad Ehlers/Photographer's Choice (front); Getty Images/Photo 24/Brand
X Pictures (back left); Getty Images/Jeff Hunter/Photographer's Choice (back middle); Getty
Images/David Fleetham/Taxi (back right).

Getty Images: pages 1 (David Fleetham/Taxi), 3 (Hans Neleman/Stone), 4 (Chad
Elhers/Photographer's Choice), 5 (Jeff Hunter/Photographer's Choice), 6 (Digital Vision), 7L
(Georgette Douwma/Digital Vision), 7R (Jane Burton/Dorling Kindersley), 11T (Digital Vision),
11B (Wayne Walton/Lonely Planet Images), 14 (Jeff Rotman/The Image Bank), 15L (Paul
Nicklen/National Geographic), 15R (Brian J. Skerry/National Geographic), 16 (Hans
Neleman/Stone), 17TL (David Fleetham/Taxi), 17BL (Georgette Douwma/Photographer's
Choice), 17TR (altrendo nature/Altrendo), 17BR (Photo 24/Brand X Pictures), 18L (Gary
Bell/Taxi), 18TR (E. Pollard/Photolink/Photodisc Green), 18BR (Carolina Biological/Visuals
Unlimited), 19 (Nikolas Konstantinou/Stone), 20L (Stephen Frink/Photographer's Choice), 20R
(David E. Meyers/Stone), 21 (Georgette Douwma/Taxi), 22 (Chick Davis/Stone), 23L (Digital
Vision), 23R (James Gritz/Photodisc Green), 24 (Ken Graham/Stone), 25 (Digital Vision), 26
(Pedro Armestre/AFP), 27L (Brian J. Skerry/National Geographic), 27R (Paul Nicklen/National
Geographic), 28-29 (Chad Ehlers/Photographer's Choice), 29T (Photo 24/Brand X Pictures),
29M (Stephen Frink/Photographer's Choice), 29B (Chad Ehlers/Photographer's Choice), 30
(Digital Vision).

Cover description: Seychelles is
an archipelago of about 115
islands located 1,000 miles
(1,609 kilometers) east of Kenya
in the Indian Ocean.

All of the Internet URLs given
in the book were valid at the
time of publication. However,
due to the dynamic nature of
the Internet, some addresses
may have changed, or sites may
have ceased to exist since
publication. While the author
and publisher regret any
inconvenience this may cause
readers, no responsibility for
any such changes can be
accepted by either the
author or the publisher.

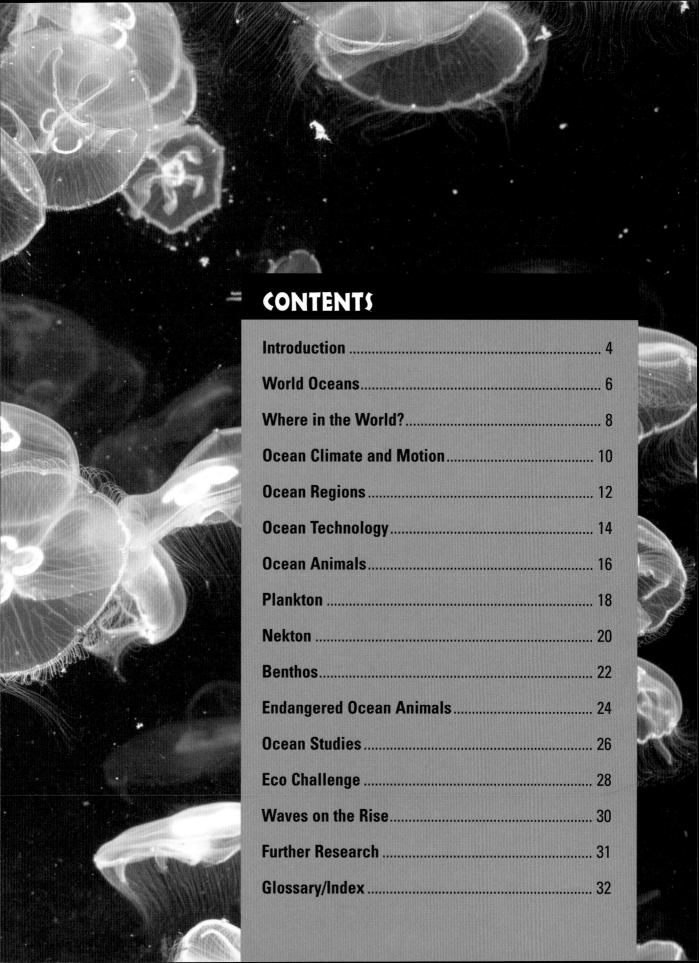

CONTENTS

Introduction

arth is home to millions of different **organisms,** all of which have specific survival needs. These organisms rely on their environment, or the place where they live, for their survival. All plants and animals have relationships with their environment. They interact with the environment itself, as well as the other plants and animals within the environment. This interaction creates an **ecosystem.**

The Indian Ocean covers about 14 percent of Earth's surface.

Different organisms have different needs. Not every animal can survive in extreme climates. Not all plants require the same amount of water. Earth is composed of many types of environments, each of which provides organisms with the living conditions they need to survive. Organisms with similar environmental needs form communities in areas that meet these needs. These areas are called biomes. A biome can have several ecosystems.

Oceans are biomes. They cover more than 70 percent of Earth's surface and supply about 97 percent of the world's water.

There are five oceans on Earth. They are the Arctic, the Atlantic, the Indian, the Pacific, and the Southern Oceans. Each ocean has smaller seas, bays, and gulfs. These oceans connect to form a large ocean. This large ocean is often called the world ocean.

For more than a century, scientists called oceanographers have studied the ocean. They have found millions of animal species and new ecosystems. Still, many ocean habitats remain unexplored.

FASCINATING FACTS

About one-third of the world's oil is drilled from oceans.

If all the gold were mined from the oceans, each person on Earth could be given 9 pounds (4.1 kilograms).

Bottlenose dolphins can live in a variety of marine habitats, from deep oceans to shallow shorelines.

World Oceans

Although they are connected, each of the five oceans exists in a distinct geographic location. The Pacific Ocean, at 63,800,000 square miles (165,200,000 square kilometers), is the world's largest ocean. The Pacific Ocean borders the western coasts of North America and South America. It also borders the eastern coasts of Asia and Australia.

Earth's second-largest body of water is the Atlantic Ocean. This 41,100,575-square-mile (106,450,000-sq-km) ocean borders the eastern coasts of North America and South America, and the western coasts of Europe and Africa.

The Indian Ocean is 28,400,130 square miles (73,556,000 sq km) in area. Bordered on the north by southern Asia, on the west by Africa, and on the east by Australia and the Sunday Islands, the Indian Ocean is the third-largest body of water in the world.

The Pacific Ocean borders the eastern coast of New South Wales in Australia.

At 7,848,299 square miles (20,327,000 sq km), the Southern Ocean, which encircles Antarctica, is the fourth-largest body of water on Earth. Prior to 2000, this ocean did not officially exist. It was considered part of the Indian, Atlantic, and Pacific Oceans.

The smallest of the five oceans, the Arctic Ocean, is 5,440,179 square miles (14,090,000 sq km) in area. Europe, North America, Asia, Greenland, and many small islands border the Arctic Ocean.

FASCINATING FACTS

The average depth of the ocean is 12,460 feet (3,800 meters). This is ten times the height of New York City's Empire State Building.

Seahorses live in temperate and tropical coastal ocean waters. They are usually found in coral reefs and seagrass beds.

Lionfish are found in the waters between Australia and Japan.

WHERE IN THE WORLD?

Oceans cover much of Earth's surface. This map shows where each of the world's oceans is located. Find the place where you live on the map. Does the place where you live border an ocean? If not, which ocean is nearest to your home? Which countries border each ocean?

Ocean Climate and Motion

O cean climates vary because oceans are located in various parts of the world, from the freezing Arctic to the warm waters of the tropics. Near polar regions, surface water temperatures dip to about 28° Fahrenheit (−2° Celsius), while warmer waters reach temperatures of 97° F (36° C). The average surface temperature of the world ocean is 63° F (17° C). However, in most cases, as the water becomes deeper, the temperature decreases. This occurs because the surface water absorbs most of the Sun's energy. The Sun's heat does not reach the deeper parts of the ocean. The thermocline, which begins between 300 and 1300 feet (100 and 400 m) below the ocean's surface, is an area where water temperature decreases greatly. In this area, temperatures drop as low as 32° F (0° C).

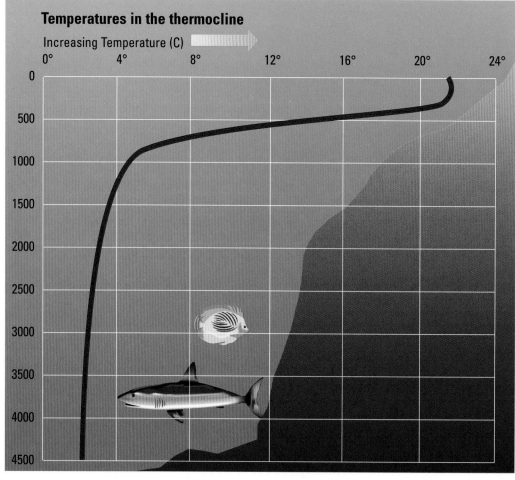

Nearly 90 percent of ocean water lies below the thermocline.

The highest point on a wave is called the crest. The lowest point is called the trough.

Ocean waters move constantly. Waves, currents, and tides all work to keep ocean water in motion.

Waves

Winds, earthquakes, volcanoes, and landslides create waves. Winds blowing across the ocean form surface waves. The size of these waves depends on how fast, far, and long the wind blows. Tsunamis, or tidal waves, are huge waves that can cause great damage on land. They are caused by underwater earthquakes, volcanoes, and landslides that create shifts in the bottom of the sea.

Currents

An ocean current is a continuous stream of water moving along a definite path. This stream can run vertical or horizontal to the water's surface. It can also be near the surface or deep below. Currents can be warm or cool. For example, water flowing away from the equator is warmer than water flowing toward the equator.

Tides

Tides are the regular rise and fall of the level of ocean water over a certain period of time. The **gravitational pull** between Earth and the Moon causes tides. Two tides occur each day.

FASCINATING FACTS

Oceans can affect temperatures on land.

As oceans warm and cool more slowly than land does, coastal areas tend to have cooler summers and warmer winters than inland areas do.

Ocean Regions

Oceans surround every continent on Earth. Coastlines bordering the world's oceans slope downward at a slight angle. The angle of the slope gradually increases—leading to deeper water. This slope is called the continental shelf. In some parts of the world, the continental shelf extends many miles into the ocean. In other areas, the shelf reaches only a short distance past the shore.

The continental shelf leads to the continental slope. The continental slope is a sharp drop where the water becomes quite deep. In most areas, the continental slope begins at a depth of 430 feet (131 m). It can extend as deep as 12 miles (20 km). Often, the continental slope becomes a smooth, gently sloping area called the continental rise. The continental rise is part of the ocean bottom. Beyond the continental rise lies the deep ocean basin.

The deep ocean basin is about 2.5 to 3.5 miles (4 to 5.6 km) deep. It covers nearly one-third of Earth's surface. The ocean's bottom, known as the abyss, is located in the deep ocean basin. Deep-sea trenches are also located in this part of the ocean.

The Land Beneath the Oceans

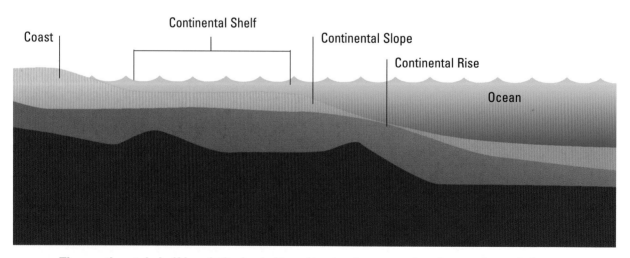

The continental shelf is relatively shallow. It extends outward to the continental slope, where the deep ocean begins.

Ocean Depth Levels

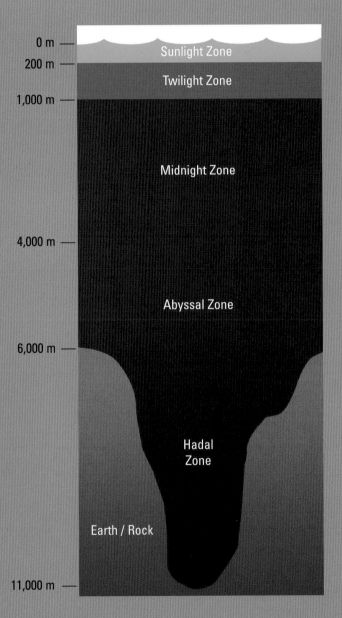

The ocean has five main zones, or depth levels, based on the level of sunlight an area receives.

The epipelagic, or sunlight, zone is the ocean's shallow, top layer. This layer extends about 656 feet (200 m) below the water's surface. Here, there is enough sunlight for plants to survive.

Located 656 to 3,281 feet (200 to 1,000 m) below the ocean's surface, the mesopelagic, or twilight, zone receives very little sunlight. Plants cannot live in this region.

Sunlight does not reach the deep ocean area known as the bathypalagic, or midnight, zone. This zone is located about 3,281 to 13,123 feet (1,000 to 4,000 m) below the water's surface.

The abyssal zone, located 13,123 to 19,685 feet (4,000 to 6,000 m) below the ocean's surface, is very dark. The water in this region is near freezing temperatures.

The ocean's deepest trenches are called the hadal zone. This zone is located between 19,685 and 36,089 feet (6,000 and 11,000 m) below the ocean's surface.

Ocean Technology

For centuries, the mysteries of the oceans have intrigued people. Explorers and early scientists studied the environment beneath the ocean's surfaces. Today's scientists continue to study oceans. They look for ways to extract and use the ocean's resources. Scientists continually develop new technologies for ocean exploration and research.

In 1822, using a basic underwater bell, Daniel Colloden discovered he could accurately measure the speed of sound underwater. Since then, the technology of underwater **acoustics**, or sonar systems, has greatly advanced. Scientists now know that sounds behave differently in unique ocean environments. Sonar causes sound waves to bounce off objects in the water. Scientists can calculate the distance between two objects in the water by measuring the time it takes for a sound to bounce between the objects. Sonar helps scientists measure the size, location, and motion of underwater objects. As a result, scientists can map the ocean and its features.

Scientists can view great white sharks close up from the safety of a shark cage.

Divers use measuring tapes to map a coral reef's area.

Scientists use a variety of special equipment to acquire information about the ocean. Since the ocean floor has never been touched, it is one of the best indicators of Earth's history. Scientists obtain sediment samples using special machines that drill into the ocean floor. This is called core drilling. Scientists study core samples to learn about ocean physics, biology, and chemistry. Core drilling has helped scientists gain valuable information about sea levels, polar icecaps, and **plate tectonics**.

Scientists often use **submersibles** to enter cold, dark ocean areas. These areas are under large amounts of pressure because they are so deep in the ocean. Using submersibles, scientists examine great depths. They can uncover new ecosystems and species. Scientists also use vehicles operated by remote controls to collect data. The vehicles collect samples and take photographs in areas people cannot access. These vehicles then send television signals to scientists aboard ships.

FASCINATING FACTS

The aqualung, an underwater breathing aid, was developed in the 1940s. This device allows people to dive to great depths without rising to the surface for air. Today, the aqualung is called SCUBA (self-contained underwater breathing apparatus).

Aquarius is located on the ocean floor off the coast of Florida in Key Largo. The underwater laboratory is home to many ocean research projects.

OCEAN ANIMALS

From whales and sharks to sea anemones and eels, Earth's oceans are teeming with life. More than 1 million plant and animal species live beneath the water's surface. Marine mammals, seaweed, fish, birds, insects, reptiles, and amphibians all make their homes in the ocean. Some live in warm, tropical waters, while others live in the murky depths of the abyss. Ocean life is divided into three groups: plankton, nekton, and benthos.

PLANKTON

Plankton are floating or drifting plants and animals. They are microscopic, meaning they cannot be seen by the naked eye. Plankton move with the water currents, waves, or tides.

Plankton plants are called phytoplankton. Phytoplanktons gain energy from saltwater minerals and sunlight. They are most numerous near shores where there is abundant sunlight. All underwater animals feed on phytoplankton or other phytoplankton-eating animals.

Plankton animals are called zooplankton. Zooplankton feed on microscopic animals and serve as food for larger underwater creatures. Examples of zooplankton include copepods, jellyfish, and arrowworms.

About ninety-eight percent of a jellyfish is made up of water.

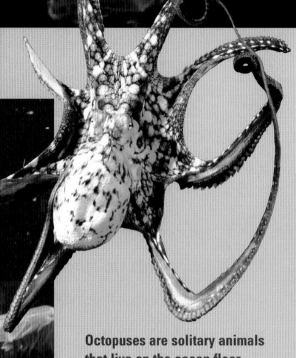

Oysters have been on Earth for millions of years. Today, they are used mainly as food.

Octopuses are solitary animals that live on the ocean floor.

NEKTON

Most underwater animals are nekton. Nekton can swim independently of flowing currents. They often live in areas where there is a specific temperature, food supply, and salt solution in the water. Examples of nekton include sharks, eels, octopuses, and other fish.

BENTHOS

Benthos are animals and plants that live on the bottom of the ocean. Some benthos animals, such as oysters, remain in one place their entire lives. Others, such as starfish and lobsters, move by swimming or walking. Benthos plants can only grow in areas of the ocean that sunlight reaches. For example, seagrasses are common benthos that grow along shorelines. All benthos plants are attached to the ocean floor, remaining in one position their entire lives.

Many colorful fish reside in the waters of the tropics.

If a starfish loses an arm, it grows another one in its place.

Plankton

Crustaceans

Small **crustaceans**, such as larval crabs and lobsters, are zooplankton. Crustaceans are **invertebrates.** There are more than 30,500 known crustacean species on Earth. They are the most populous species in the ocean. Although most crustaceans are very small, they come in many sizes, colors, and shapes. Crustaceans have three body parts: the head, the **abdomen**, and the **thorax**. Crustaceans do not have bones. Instead, all crustaceans have a hard **exoskeleton.** They also have a **segmented** body. This means that the right side of the body is identical to the left side of the body. Crustaceans have two pairs of antennae that they use to touch and smell. They have jointed legs, which they use to swim or walk. They also have a pair of eyes and a pair of mandibles, or jaws.

Lobsters live in the ocean's crevices and caves.

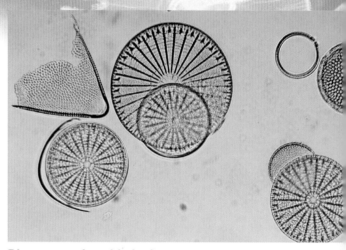

Diatoms are found in both salt and fresh waters.

Diatoms

Diatoms are single-celled plants that float through water. They have a hard shell that sinks when the plant dies. There are more than 10,000 diatom species. The largest individual diatoms can barely be seen by the naked eye. Diatoms do not have roots, stems, or leaves. They contain chlorophyll, which absorbs sunlight and turns it into food energy. Diatoms often form colonies.

Green algae

Green algae are single-celled plants that contain chlorophyll, which they use to capture sunlight and create food energy. There are more than 800 ocean species of green algae. Most green algae species live in fresh water. About 10 percent live in ocean environments.

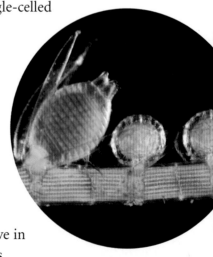

There are more than 7,000 species of green algae.

Jellyfish

Many types of jellyfish are zooplankton. Like crustaceans, jellyfish are invertebrates. They are **gelatinous** animals with soft bodies. Jellyfish have long tentacles that contain a poisonous venom, which they use to catch prey. There are nearly 3,000 known jellyfish species. Jellyfish do not have brains. They have a basic nervous system that senses light and scents. Although jellyfish come in many shapes and sizes, most are bell shaped and **transparent**. They can live in many ocean environments, ranging from shallow coastal waters to depths of 12,000 feet (3,650 m).

Jellyfish inhabit every major ocean area of the world. They can withstand a wide range of temperatures.

FASCINATING FACTS

Green algae first appeared more than 2 billion years ago, making them one of Earth's oldest organisms.

The mushroom jellyfish is a gourmet meal in China and Japan. It is served fresh or pickled.

The smallest diatoms are less than 0.0001 inch (0.0003 centimeters) long.

The largest crustacean species is the Japanese spider crab, which can have a leg span of nearly 13 feet (4 m), a body size up to 15 inches (38 cm), and a weight up to 44 pounds (20 kg).

Pea crabs are the smallest crustaceans. They are about 0.25 inches (0.64 cm) across the shell.

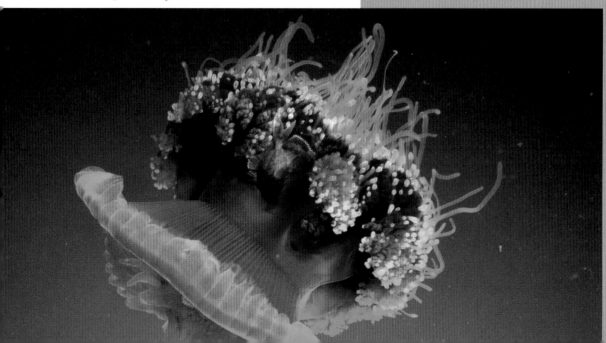

Nekton

Sharks

More than 375 shark species live on Earth. Sharks are found in every ocean on Earth. Some live in tropical waters. Others live in arctic regions. Sharks can live in shallow or deep waters. They eat almost any type of animal they find. Most sharks have triangular teeth to catch, hold, and tear prey. When catching prey, a shark's upper jaw extends outward. The lower teeth catch and hold the prey, while the upper teeth bite through the animal. Sharks are cartilaginous fish. This means that their skeletons are made from cartilage, an elastic-like tissue. Ranging in length from 8 inches (20 cm) to 14 feet (4 m), sharks' **streamlined** bodies are rounded

Orcas live in all of the world's oceans, but are most numerous in colder waters.

in the center, coming to a point at either end. Sharks have five fins that help them move through water.

Marine Mammals

Marine mammals are warm-blooded animals that are covered with hair or fur and breathe through lungs. Examples include dolphins, whales, manatees, dugongs, seals, walruses, sea lions, polar bears, and sea otters. Marine mammals live in all of the world's oceans. Most marine mammals have a thick layer of fat, called blubber, between the skin and muscle. Blubber keeps these animals warm in cold waters. Manatees have only a thin layer of blubber. They live in tropical and subtropical waters no cooler than 70° F (21° C). Most marine mammals eat fish, squid, shellfish, or even other marine mammals. Sirenians, a group that includes manatees and dugongs, are the only herbivores, or plant-eaters.

Great white sharks are usually found close to the water's surface and in coastal waters, where their prey lives.

Sea Turtles

There are seven sea turtle species. A sea turtle is a large, cold-blooded animal with four flipper-like limbs that it uses to propel through water and walk on land. A turtle also has a shell attached to its back. The top of this shell, called the carapace, ranges in color, length, shape, and scales. Sea turtles range in size from 2 to 6 feet (1 to 2 m) long. Sea turtles live in tropical and temperate waters. In their early years, sea turtles live near shore. They spend the remainder of their lives in the open ocean. During the breeding season, they may **migrate** to their nesting grounds, which are located at the same place the turtle hatched.

Most sea turtles are carnivores, eating crabs, shrimps, lobsters, small fish, and jellyfish. Sea turtles do not have teeth. Their jaws are adapted to eat certain foods. The hawksbill has a beak, which it uses to find food in crevices and coral reefs. Adult green turtles are the only herbivorous sea turtles. Their jaws have serrated edges to tear apart plant materials. Other sea turtles have powerful jaws that crush prey.

Although green sea turtles spend most of their lives in the ocean, females come to shore to lay their eggs.

FASCINATING FACTS

The marine iguana lives on the Galapagos Islands. It is the only lizard that swims in the ocean.

Sea turtles lay up to 200 eggs at one time. Only a few hatchlings will survive predator attacks to reach maturity.

At up to 46 feet (14 m) long, the whale shark is the largest shark. It is also the largest fish.

Adult manatees eat 5 to 10 percent of their body weight each day.

Benthos

Sea Anemones

Sea anemones are invertebrates that come in many shapes, sizes, and colors. There are more than 1,000 species of sea anemones. Most are small—reaching only 1 to 4 inches (3 to 10 cm) across. Others can grow up to 5 feet (2 m) across. Each sea anemone has a vase-shaped body. Stinging tentacles surrounding the mouth protect sea anemones from predators. Clown fish, which are **immune** to the anemone's sting, eat decomposed matter from sea anemones' tentacles.

Sea anemones can live in any part of the ocean. They are most common in tropical waters, where they attach to the ocean floor, reefs, and corals, or burrow beneath the sand and mud. Sea anemones cannot hunt their food. They wait for prey to swim nearby before stinging it with their poisonous tentacles. These tentacles move the food into the sea anemone's mouth. Sea anemones eat fish, mussels, crustaceans, and **larvae**.

A sea anemone spends most of its life in one place.

Seagrasses

Seagrasses grow in shallow waters where they receive large amounts of sunlight. Like land plants, seagrasses have roots, leaves, seeds, and flowers. They use their roots to anchor themselves to mud and sand on the ocean floor. From the blade-like eelgrass that grows up to 13 feet (4 m) long to the rounded sea vine that grows about 0.8 to 1.2 inches (2 to 3 cm) long, about 60 species of seagrasses live in the ocean. Seagrass is an important food source for other ocean animals. It provides habitat and nutrients, and prevents **sediments** from moving.

If a starfish loses an arm, it can grow a new one.

There are currently fifty-eight species of seagrasses. Australia is home to at least thirty of the species.

Cleaner shrimp are immune to the sting of sea anemone tentacles. They hide from predators inside the tentacles, while eating decaying matter off the sea anemone.

The blue starfish can be spotted in the Great Barrier Reef on Australia's coast.

Starfish

More than 1,800 species of starfish live in Earth's oceans. The largest variety of species live in the Pacific Ocean. Most of these invertebrate animals have five hollow arms covered with spines on top and rows of tiny, tube-like feet on the bottom. These "feet" have suction cups, which the starfish use to slowly move and grab onto objects. Most starfish are about 8 to 12 inches (20 to 30 cm) across. They can range in size, however, from 0.4 inches (1 cm) to 26 inches (65 cm). Starfish eat a variety of animals, including mussels and mollusks.

Endangered Ocean Animals

Animals in danger of becoming extinct are classified as endangered. This means there are so few of the species alive that they need protection to survive. The blue whale, hawksbill turtle, smalltooth sawfish, and shortnose sturgeon are just a few of the animals considered to be endangered. In the United States, people are not allowed to hunt or harm endangered animals.

Human destruction of habitats and pollution threaten the world ocean. Pollution increases nitrogen gas levels in the water, causing large amounts of algae to grow. In coral reefs, this is particularly dangerous. Excessive algae prevents necessary sunlight from reaching the reefs—killing ocean habitat and the fish dependent on it for survival.

Water currents can carry garbage dumped on ocean coastlines thousands of miles into the sea.

In other cases, overharvesting endangers certain underwater animals. Overharvesting means that too many animals are hunted. When this happens, there are not enough animals left to maintain the population. Shark fins, meat, and livers are in great demand for food or health and beauty ingredients, putting many shark species at risk of becoming endangered. Southern sea otters were once overhunted for their furs. Today, these animals are protected from such overharvesting.

Between the late 1950s and the 1990s, about 6 million dolphins died in fishing nets.

Many marine species become entangled in tuna fishing nets. Throughout the 1970s and 1980s, more than 100,000 dolphins died in tuna nets each year. Today, fishers use special fishing methods to reduce the harm caused to ocean animals.

OCEAN STUDIES

From working with whales to guiding submersible vehicles through the abyss, or from developing research devices to collecting core samples, marine careers are exciting and challenging. Marine careers require a background in math, science, and computer courses. Before considering a marine career, it is important to research career options and visit marine centers that offer hands-on experiences.

MARINE BIOLOGIST

- Duties: studies ocean life and how ocean organisms interact with their environments

- Education: bachelor's, master's, or doctoral degree in marine science

- Interests: math, science, oceanography, animals

Marine biologists enjoy learning about organisms and their environments. To understand how animals and plants interact with their environments, marine biologists must also study chemical, physical, and geological oceanography. The ocean is filled with many different creatures, so most marine biologists choose to study a specific subject.

OCEAN ENGINEER

- Duties: develops tools to study the ocean

- Education: engineering degree

- Interests: math, physics, chemistry, mechanics

Ocean engineers develop the tools that marine biologists and oceanographers use to study the ocean and its life. Satellite-linked buoys, sediment traps, underwater video equipment, acoustic measuring devices, and underwater vehicles are just a few of the tools ocean engineers have developed.

OCEANOGRAPHER

- Duties: studies the geology and geography of the ocean

- Education: bachelor of science degree; master's degree in oceanography

- Interests: physics, geology, biology, chemistry, geography

There are many types of oceanographers. Each specializes in a different part of the ocean. For example, physical oceanographers study the relationship between the sea and the atmosphere. Geological oceanographers study the rocks, sediment, and other features of the ocean floor.

ECO CHALLENGE

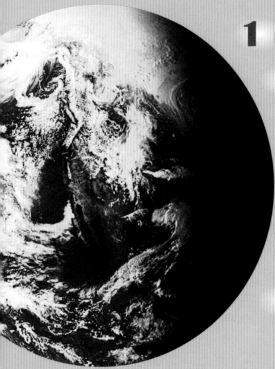

1 How many oceans does Earth have? Name them.

2 What three things make water move?

3 What is the continental shelf?

4 In which zone are the ocean's deepest trenches located?

5 What is the world ocean's average surface temperature?

6 All ocean animals belong to one of three groups. Name these groups.

7 What type of fish are sharks?

8 What three body parts do all crustaceans have?

9 How many sea turtle species live on Earth today?

10 Name three ways that ocean species become endangered.

Answers

1. five; Arctic, Atlantic, Indian, Pacific, Southern
2. waves, currents, and tides
3. The world's oceans slope downward at a slight angle. The slope gradually becomes greater—leading to deeper water. This slope is called the continental shelf.
4. hadal zone
5. 63° F (17° C)
6. plankton, nekton, and benthos
7. cartilaginous fish
8. head, abdomen, and thorax
9. seven
10. pollution, habitat destruction, overharvesting, fishing nets

WAVES ON THE RISE

Try the following experiment to learn how wind speed affects the height of waves and how higher waves occur in shallower water.

MATERIALS

- bendable straw
- 9 x 13-inch (23 x 33-cm) baking dish
- duct tape
- water
- felt pen
- ruler

1. Bend the straw so that it forms an "L."

2. Position the straw inside the baking dish so the short end is facing upward. The long end should be about 0.5 inch (1.2 cm) above the bottom of the dish.

3. Pour water into the baking dish until it reaches just below the straw.

4. Place duct tape on side of baking dish to record "wave" measurements.

5. Blow gently into the short end of the straw. Using the felt pen, mark the wave height on the outside of the dish.

6. Repeat two more times, blowing harder each time.

7. Remove the water from the dish, and reposition the straw so that it is nearer the top of the dish.

8. Pour water into the dish until it reaches just below the straw, and repeat the experiment.

9. Observe the "wave" heights based on the "wind" speed and water depth.

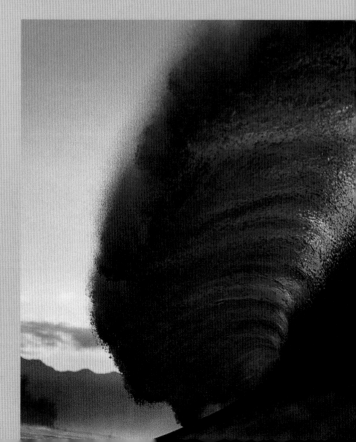

FURTHER RESEARCH

How can I find more information about ecosystems, oceans, and animals?

- Libraries have many interesting books about ecosystems, oceans, and animals.

- Science centers and aquariums are great places to learn about ecosystems, oceans, and animals.

- The Internet offers some great Web sites dedicated to ecosystems, oceans, and animals.

BOOKS

Earle, Sylvia. *Atlas of the Ocean: The Deep Frontier*. Hanover, PA: National Geographic, 2001.

Littlefield, Cindy A. and Sarah Rankin. *Awesome Ocean Science: Investigating the Secrets of the Underwater World*. Nashville, TN: Williamson Publishing Company, 2002.

Prager, Ellen J. and Sylvia Earle. *The Oceans*. Columbus, OH: McGraw Hill, 2001.

WEB SITES

Where can I find a good reference Web site to learn more about oceans and animals?

The Ocean Channel
www.ocean.com

How can I learn more about oceans?

The Ocean Conservancy
www.oceanconservancy.org

Where can I learn more about the thousands of animals that live in the oceans?

SeaWorld Animal Information
www.seaworld.org/animal-info/index.htm

GLOSSARY

abdomen: the part of the body where the digestive organs are located

acoustics: designed to carry sound

crustaceans: animals with segmented bodies, jointed limbs, and an outer shell

ecosystem: a community of living things sharing an environment

exoskeleton: an outer covering

gelatinous: resembling jelly

gravitational pull: the pulling force of mutual attraction between the planets, stars, or particles

immune: able to resist

invertebrates: animals that do not have backbones

larvae: newly hatched animals

migrate: to move from one place to another

organisms: individual life forms

plate tectonics: layers of Earth's crust that move and float

sediments: materials that settle to the bottom of a liquid

segmented: divided into parts

streamlined: designed to move easily through water or air

submersibles: small submarines

thorax: second part of a crustacean's body

transparent: allowing light to shine through

INDEX